Simulation Series

A Series of Simulations Designed to Challenge High-Ability Learners in History, English, and the Humanities

Earth Friendly

By Charlene Beeler, M.S.Ed.

Prufrock
Press
Post Office Box 8813
Waco, Texas 76714-8813
1-800-998-2208

Table of Contents

Forward

When Charlene Beeler submitted her simulations to me, I was excited. This teacher of almost 20 years had designed a set of four units which were highly appropriate for high-ability learners in the secondary classroom. The units, I felt, could be easily used as models for teachers seeking fresh ideas and methods for teaching high-ability learners.

Of course, it should be noted, that the units within the *Simulation Series* offer examples of one strategy for teaching high-ability youngsters—a simulation. There are many others. The series was not designed to offer *the strategy* for teaching these students—it simply acts as an example of one way.

All of the units within the series are designed to be modified by you *before* they are used in the classroom. These units are not the "anybody-can-teach-em" units I so often encountered as a classroom teacher. These units act as curricular shells which should be modified to align with your personal objectives for your students and the curricular objectives of your district. While preparing for this unit, you will want to add objectives and tasks and develop appropriate evaluation tools.

While you use these units in your classroom, your role will be that of a facilitator or resource person. You will help students locate information and materials. You may also act as the evaluator of student products. However, your importance in this latter role may vary depending on the level of independence your students have gained.

Most importantly, enjoy the units in the series. The excitement and scholarship they promote in the classroom is a real pleasure to observe.

—Joel E. McIntosh
Editor of *The Journal of Secondary Gifted Education*

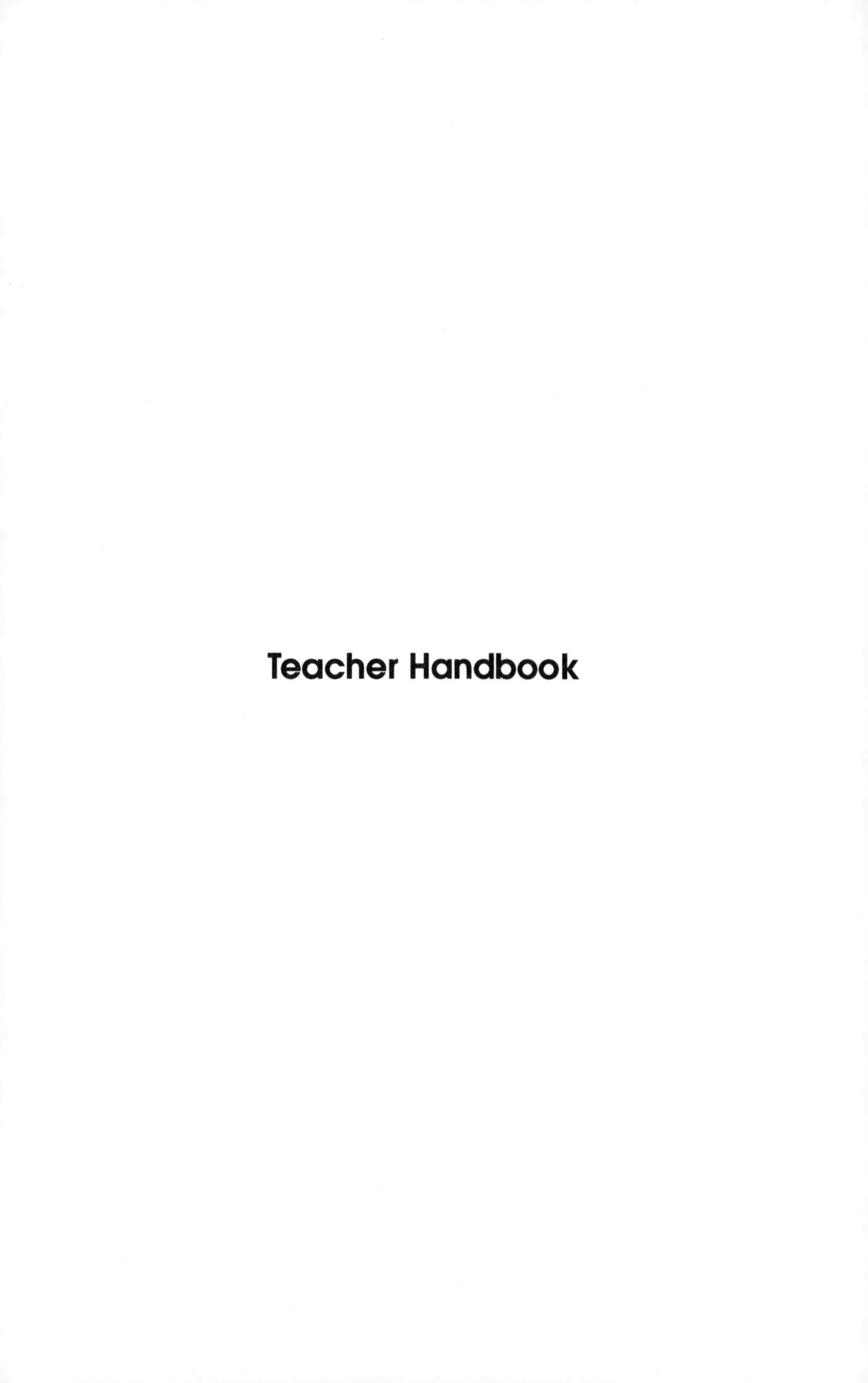

Teacher Handbook

INTRODUCTION

You are about to participate in a cooperative learning project. Each member of each group works toward the same goal—that of helping the group accomplish the end result.

Each team of four people is responsible for developing "The Perfect Earth Friendly City." They will strive to win points that carry them to the ultimate in living: a place where land provides a safe dwelling for all wildlife, lush foliage surrounds all inhabitants, air is free of pollutants, and toxic waste and nuclear disasters threaten no one.

Part of the project requires students to simulate the personality of make-believe characters. As this character, they will be given responsibilities and tasks to accomplish, a problem to solve and decisions to make. Afterwards they will create a visual art statement projecting their opinions concerning ecology. In conclusion, they will write an essay detailing what they have learned about our world today and provide some thought on it and how they will live their lives differently in the future.

The way the project progresses depends upon how many EPs (Environmental Points) each team accumulates. EPs propel the teams forward on the map as they travel along their journeys to "The Perfect Earth Friendly City." The group that accumulates the most EPs wins the simulation and the award of "Perfect Citizens" will be theirs.

The following is what the students will accomplish during simulation play:

1. interact with a group
2. analyze and make decisions
3. create two-dimensional or three-dimensional art work
4. come to an agreement on solutions about a crisis intervention plan
5. compete with other groups for points
6. learn about dangers to the Earth
7. verbalize information, concerns and possible solutions in a presentation
8. research a problem and come to a conclusion about their futures
9. present their opinions to others
10. evaluate their work

Of Time and Grades

You may shorten or lengthen the unit by eliminating or adding different activities.

1. Introduction to the simulation
2. Video about the environment—suggestions for video:
   "Exxon Valdez" (an HBO special)
   "The China Syndrome"
   "Silkwood"
3. Groups formalize their roles and begin the project which includes:
   a. ecology test
   b. individual research work
   c. crisis prevention plan
   d. art work
   e. thought questions
   f. essay test
   g. presentation of work
   h. debriefing

Approximate total hours of play = 18

All reports, presentations, and projects will be graded according to quality, length, and creativity. The chart provided below will help you award EPs.

50 EPs = outstanding quality effort
40 EPs = good quality effort
30 EPs = above average effort
20 EPs = average effort
10 EPs = unsatisfactory effort

TEACHER PREPARATION

1. Read Introduction beginning on page 8.
2. Become familiar with the student packet beginning on page 13.
3. Make copies of necessary student materials.
4. Make a poster-sized board for the "Perfect City." (suggested map on next page)
5. Prepare for facilitation by reading Teacher Strategy on page 12.
6. Rent or check out an introductory video from the library or video store.
7. Preview the video; prepare questions involving the video for students to answer.
8. Check out reference pictures on toxic waste, spills, DDT, and other pesticides, over-population, industrial waste, etc., and display them in your classroom.

Materials Needed

1. Poster board and tape for constructing a "Journey to the Perfect City" (suggestion for drawing map included on next page)
2. Markers
3. Push pins and tags for identification purposes to attach to the map
4. Reference books on toxic wastes and spills
5. Movie or video
6. Student packets
7. Art supplies

Journey to the Perfect City

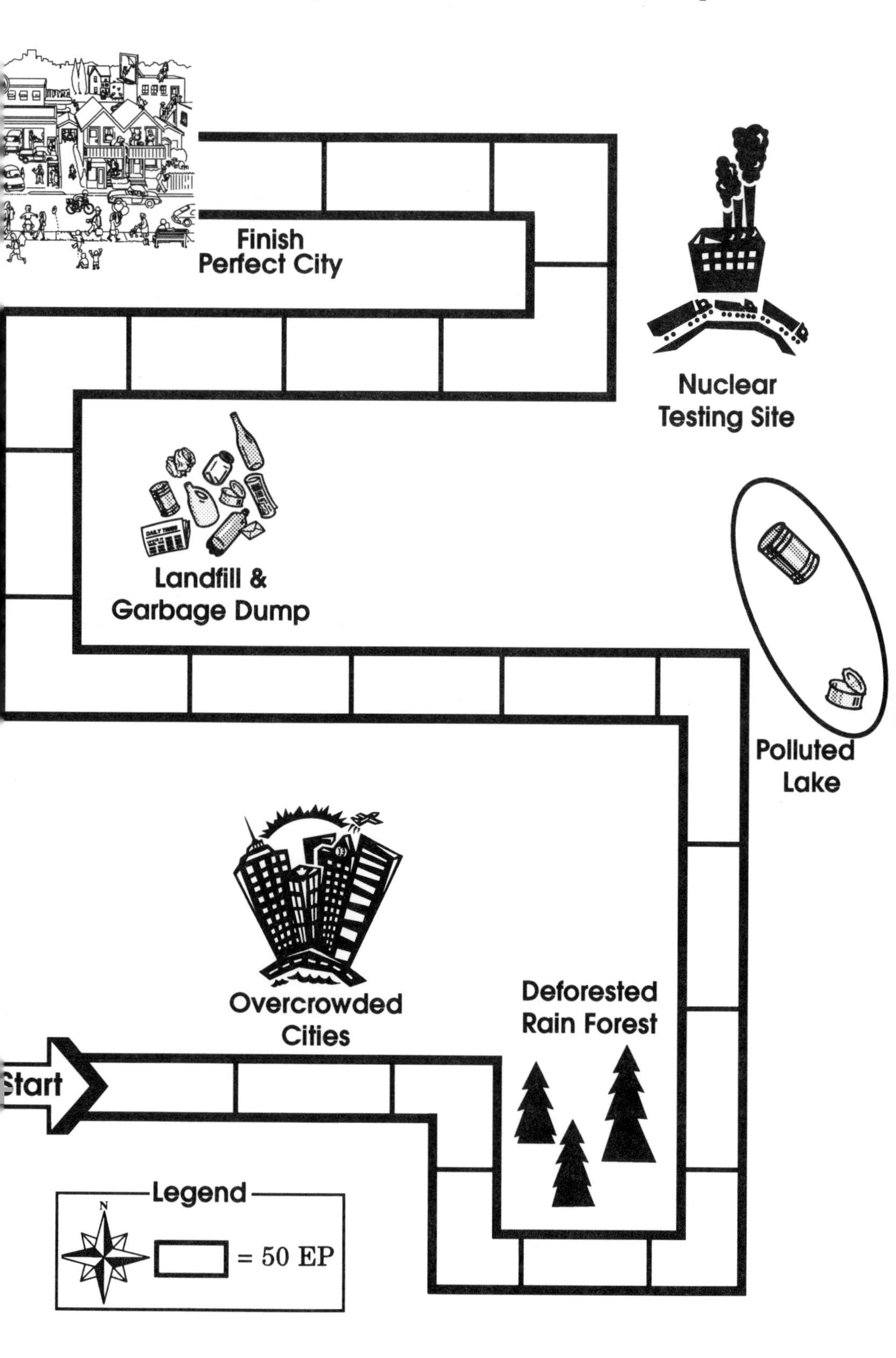

Teacher Strategy

1. Gather students together.

2. Introduce "Earth Friendly" by allowing students to brainstorm different ecological problems facing our planet today. Write their ideas on the board.

3. Point out that although there are many areas of concern (as shown by the number of responses in the brainstorming activity), the class can only focus on a few, mainly toxic and industrial waste, pesticides, global warming, and deforestation. Endangered animals are addressed in a separate simulation project called "Endangered Species."

4. Pass out Student Packets.

5. Ask a student to read the Introduction aloud.

6. Introduce the map, "Journey to the Perfect City," and explain to the students that this is the way their groups will keep track of their EPs. (The mayor of each group will advance the group's push pin according to how many EPs have been awarded. The work and quality of the work will determine how many EPs are awarded.)

7. Show the video at this time, having briefed students on what specifically to listen for.

8. After showing the video, divide the class into groups of four.

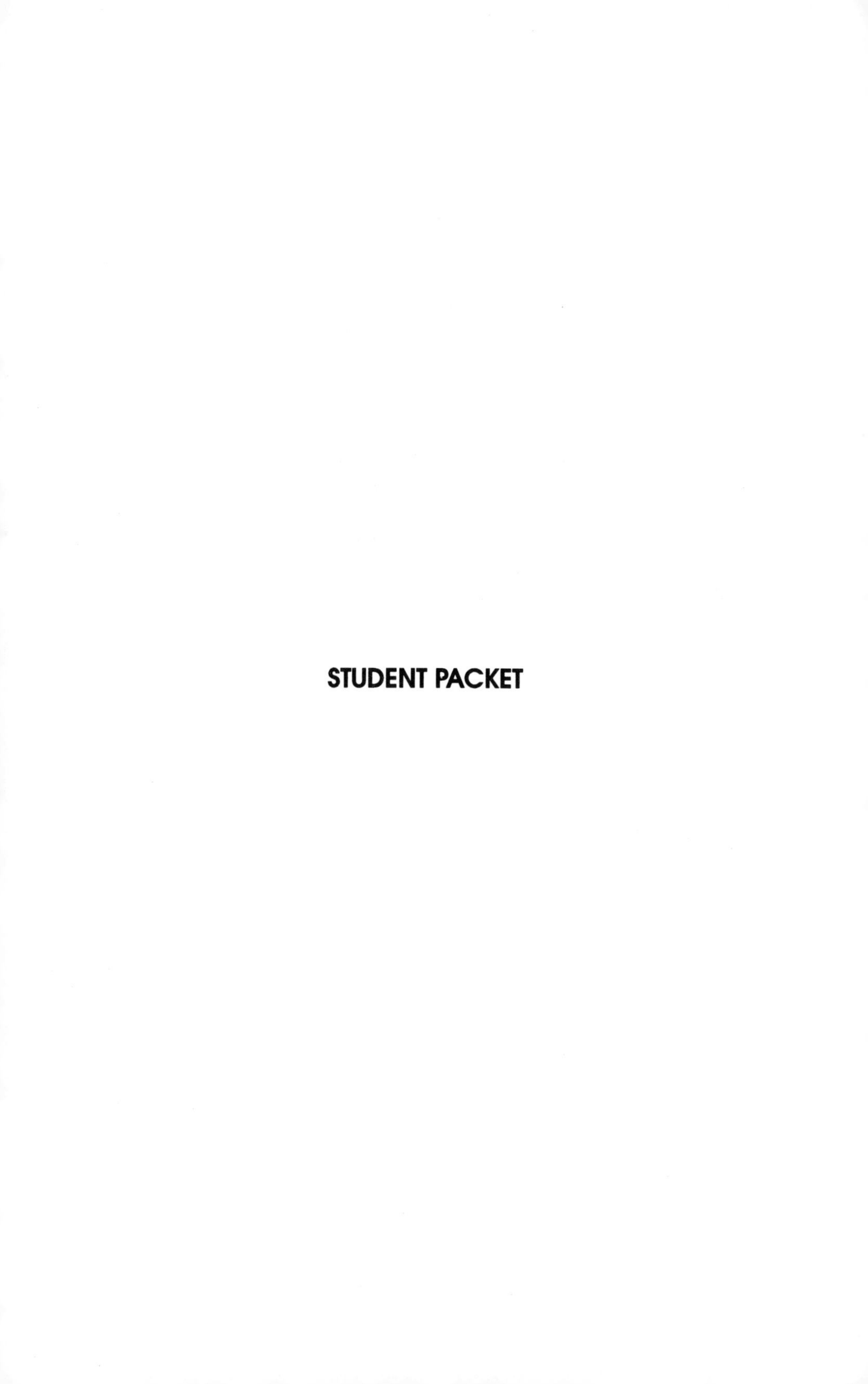

STUDENT PACKET

TIME CHART

DAY 1

1. Select groups.
2. Help group decide on a name for your "Perfect Earth Friendly City." Write name on a thin sheet of colored construction paper provided by your teacher.
3. Have a representative from your team (probably the mayor) select a push pin and position this and the colored construction paper with your group's name written on it at the beginning of your journey marked START on the map.

One way to earn extra EPs is to find current articles in the newspaper about ecology problems or solutions and:

1. cut them out and paste them on clean, white paper;
2. summarize the articles in your own words;
3. type summary for 10 extra EPs;
4. turn in for EPs.

DAY 2

1. Turn to page 22 in your packet and read the role of characters and responsibilities.
2. Select a role.
3. Become familiar with the task assigned each role.

Read all roles and be prepared for anything. When you are thinking about which role to chose, think about your own strengths. Some of the roles require more speaking than writing, some require more researching than speaking. Talk about these things with your group and decide who is best suited for each role.

DAY 3

1. The recorder makes a Tally Chart for the purpose of adding up all EPs when awarded. Each mayor will move his or her group's push pin each day after the teacher has handed back graded work.
2. All group members take the Ecology test on page 27. The teacher will provide the mayor with test answers after the test is completed. Group members will score their own tests as the mayor calls out the answers. Each person should write in correct answers for future reference and a final test.
3. After grading your tests, go to the library and find articles to complete your tasks. Complete these tasks by the beginning of Day 5.

DAY 5

1. Read over your report so that you are familiar enough with it to TELL, NOT READ, to the rest of your group. Relate your information like you would like to hear it. Remember, all the articles being researched were famous cases, items that you should be aware of because they played a significant part in world history. Knowing about such events should help prevent such things from ever happening again.
2. After listening to the reports, each member should read the questions on page 29 and discuss the answers as a group before the reporter writes answers to be handed in for EPs.

DAY 5-7

1. The mayors will present their evacuation crisis prevention plans today. This plan could help save lives in an emergency.
2. Pay close attention to the mayor's evacuation crisis prevention plan. Comment on the good and bad points before reading the "Town in Jeopardy Crisis Prevention Plan" on page 30. You will find a problem to solve, facts to consider, a map of your town, and questions to answer.

DAY 8

1. Today you will hear presentations of the plan. Listen carefully to each plan. They may influence you and help you decide what to do for your Final Say Group Project.
2. Each plan will have strengths and weaknesses. Think about those before you decide what to do in your town when an emergency arises.

DAY 9-13

1. After reading "Cause and Effect" statements on page 35 for the purpose of gaining an insight and perhaps an idea for a visual statement, read "Art Project Suggestions" on page 37.

Note—art work is visual. It attracts attention and has a lot of impact. Use your talent to produce a statement that expresses your concern about the ecology.

DAY 14-15

1. Read "Dilemmas of Our Age" statements on page 38. For extra EPs, write an advertising campaign in response to one of the dilemmas.
2. Review the dilemma statements before thinking and planning out your "Final Say Paper" on what you have learned while participating in this project. Add some creative problem solving to your paper. (Instructions on page 40.)

Character Roles and Responsibilities

Joe or Joanne Filibuster—Mayor, sheriff, and group leader. Owns a 10,000 acre wheat farm just outside of town. Is very civic minded. Although some people say Filibuster is pushy, Joe/Joanne thinks of him- or herself as a good leader. Special interests are the land, and keeping the town free of crime. Has certainly done his or her part in tracking down dangerous criminals and testifying against them in trials. These criminals have been detained in the prison at the edge of town.

"I'll shoot any prisoner comin' onto my land," Filibuster says.

Has voted democratic in the past, but campaigned for Ross Perot in the '92 presidential election because "the man had guts." Filibuster is not married, drives a red Ford truck and loves to fish and hunt. He or she is 50 years old.

Has a bad habit of interrupting others when excited about an issue. Filibuster would do anything for you, but you would be indebted for life. In other words, he or she will make you a deal you'll probably be sorry for later. Watch out for Filibuster's terms.

Responsibilities:

1. The mayor oversees all work in the group and is responsible for keeping everyone on task. The mayor presents solutions on Presentation Day and heads all discussions in group. This person arbitrates during discussions. If no one can come to an agreement, the mayor makes a deal.
2. At the appropriate time, the mayor will go to the library and find an article on "Chernoble" in the periodicals catalog. The mayor will then summarize his or her findings and present the report to the group as Joe or Joanne. Refer to the Time Chart for the date of this presentation.

Janis or James Greenhorn—Wildlife advocate. Janis/James owns land for endangered species of animals and speaks out for the rights of animals whenever possible. Greenhorn believes that animals are our future because they help with the food chain, provide love and companionship, and nurture a land already filled with machines and industrial waste. This advocate thinks before speaking and is considered level headed. To Greenhorn, animals come first, even before some people—especially the criminals on the outside of town. Janis/James is liberal, Democratic, a hard worker and a self-starter. Is also very motivated when the subject of conversation turns to rain forests, animal rights, or disturbing the natural order of life. Has picketed against pollution at the industrial plant in town and has many enemies there.

Responsibilities:

1. At the appropriate time, this character will locate reference material on pesticides and other chemicals that have an adverse effect on nature and animals such as Agent Orange, the defoliant used in Vietnam in the late '60s and early '70s. What specifically did this chemical do to the land, the animals and people that it came in contact with? What was its initial purpose and what lawsuits have been brought against the government for using it? This person will summarize the article in his or her own words and present the findings to the group as Janis/James would. Refer to the Time Chart for the date to report. Turn in the report for EPs.
2. This person will act as the animal rights advocate during the Town in Jeopardy Crisis Prevention Plan. This person will state views clearly and persuasively.

Will or Wanda Knowles—Will/Wanda is very concerned with toxic waste, its disposal and transportation. Knowles is an advocate of clean air, the future of our children, and is most concerned with the lack of public concern, especially by the industrial community about toxic waste. Industries that pollute the rivers and drinking supply are this person's main target. Will/Wanda considers them greedy people who overlook safety for the advance of a dollar, and will speak out against them whenever possible.

This person is 26 years old, married, and has one child. Travels to the next town for work as a computer programmer. Knowles came back to the town in which he/she grew up because it represented stability. This person hates change, is somewhat stuffy, and would rather play chess with other intellectuals than socialize. This job on the town council is a responsibility, not especially a joy. Would really rather lobby the state capital to improve the welfare of the state. Has thought about running for a political office, but can't charm the people like needed to win support.

During a crisis Knowles thinks logically, as he/she is a left-brained person. This person disagrees with anything out of the ordinary, especially if it would mean altering the norm. Some of the citizens think that Knowles is strange because he/she has little patience for people who think differently. This person has been accused of being without imagination.

Responsibilities:

1. This person must locate an article in a past periodical about "The Love Canal," summarize the facts in his or her own words, and write up the report. This character will present this to the group as Will, then turn in for EPs.

Anna or Andy Salavatore—Group recorder and Public Relations Expert. Salavatore is a vivacious person, full of life and laughter. This person is an innovator of new ideas and is blessed with patience, leadership abilities, and a genuine gift for perception. Very few people make Anna/Andy angry, but the mayor comes closest to it.

Somehow this public relations expert has always managed to bring a truce to a heated situation. Even in high school Salavatore was a leader, mainly because she/he was never pushy and did not pursue the lead. This came naturally out of a concern for others and the outcome of a worthy cause.

This person is soft-spoken, and always has something positive to say. People listen. Salavatore's secret lies in listening to both sides before proposing a good compromise.

This person loves children, the elderly, animals, and those who can't speak for themselves very well. Always puts self last, even if it means she/he will be in a dangerous situation. This person works as a nurse, but is torn between studying to be a pediatrician or a veterinarian.

Salavatore is 23 years old, unmarried but engaged to the town preacher.

Responsibilities:

1. This person will record all solutions to the problem-solving tasks and have them available for the mayor to see and report on during Presentation Day.
2. This person must go to the library and locate an article or articles on dumping grounds for garbage. This person must find out if rules are being broken and the results of wastes seeping into the water supply near communities. This person should come to his or her own conclusions about what solutions can be found, write a report in his or her own words, share it with the group on the appropriate day then turn it in for EPs.

Don or Donna Banes—warden of the prison located just five miles out of town. Banes has liberal ideas, and is expert in the humane treatment of prisoners This person is very understanding with prisoners' complaints. He/she believes in fair play for everyone, and is an advocate of prison reform. Banes credits success to the availability of good programs, prison reform and dedication. No one has ever escaped from the prison.

Prison reforms, the welfare of prisoners and gaining recognition for good work is tantamount to everything for this person. In fact, he/she is thinking about running for a Senate seat and takes pride in being a politician. Banes does have the charm to go along with the job, often sweet talking people into doing things for the prison that they would never dream of doing otherwise.

This person is not married but dates all the good-looking people in town, whether they are 20 or 35. Banes lives in town but is looking at a ranch 10 miles north of town.

Responsibilities:

1. This person will call city hall and locate a source who can provide information on a crisis prevention plan. Nearly every community has drafted one or knows about one. This person must ask questions, find out what the general particulars of the plan are, and ask for advice on preparing an evacuation plan. This character will share this information with the class at the appropriate time and turn in the project for EPs.

TEST YOUR ECOLOGICAL KNOWLEDGE

1. Temperatures throughout the world during 1990 were the warmest ever recorded.
   True or False

2. The United States government allows most aerosol or spray can products, such as hair spray and deodorants, to contain ingredients called CFCs which are known to damage the Earth's atmosphere.
   True or False

3. Sulfur dioxide is a chemical that produces what is called "acid rain." A large part of the nation's sulfur dioxide pollution comes from midwestern states. Which one of the following do you think is most affected by acid rain?
   a. black bears in the Smoky Mountains
   b. school children in the midwest
   c. fish in New England streams
   d. rain forests in Central America

4. About half the world's plants, animals, and insects live in rain forests.
   True or False

5. You may have heard that the numbers of American ducks and geese have been decreasing over the past decade. The reason for this is:
   a. disease and germs
   b. loss of wilderness areas in which to live
   c. hunters
   d. air pollution

6. In 1989, the Exxon Valdez, a ship carrying oil, spilled 10 to 11 million gallons of oil near the shores of Alaska. Compared to the amount spilled by the Valdez, how much used motor oil from cars do people in the U.S. dump into drains and sewers each year?
   a. less than one tenth
   b. twice as much
   c. more than 10 times as much

7. Modern sewer systems have eliminated human waste pollution of drinking water.
   True or False

8. Most of the garbage thrown away in this country ends up buried in the ground in landfills. Which of the following takes up the most space in landfills?
   a. paper and paper products
   b. metals
   c. food scraps

9. Most of the bio-degradable packaging materials that decompose naturally break down within 10 years.
   True or False

THOUGHT QUESTIONS

1. Do you think that large companies take advantage of people? How and why or why not?

2. If you think large conglomerations take advantage of others, what do you perceive as the main cause?

3. What is the effect on the public of industrial companies that have no regard for environmental issues or a government that does not research these issues enough?

4. Do you know of anyone personally who has been affected by a large company? Who and what happened?

5. What can you do as a citizen or a student to help prevent these events from happening?

TOWN IN JEOPARDY CRISIS PREVENTION PLAN

Instructions for role play:
The mayor is in charge, but everyone has a part. After reading the "Problem to Solve," read "Facts to Consider," and "Questions to Answer."

Problem to Solve

As the town council, you are most concerned with the nuclear processing plant just 75 miles north of your community. You are all aware of how dangerous the conditions would be if there were a leak or a melt down, or if any radio active substances were leaked during transport. A strong wind could carry deadly fumes toward your town at an alarming rate.

You must be prepared. Now is the time to draft a crisis prevention plan that will suffice during such an occurrence. Your main concern should be how to best evacuate the people of the community in an orderly manner; however, evacuation will not be possible for everyone because of the unique characteristics that make up the population and terrain in your area.

Your town, Clarksville, is a resort community because of the lake and the mountains. It also serves as a retirement area for many older people. The hospital in the center of town is renowned for its treatment of burn victims. These people cannot be moved easily.

There is also a prison located 10 miles north of town. Several hundred inmates are housed there. Some of the town people will be concerned about the prisoners being allowed to leave the confines of the prison walls.

The tough thing about planning an evacuation is working through the agendas of the people on the council. Everyone has their own interests. Some compromises must be made for the good of the community. This group situation will give you the opportunity to see first-hand how people think and act together to resolve a problem before it happens.

Facts to Consider

1. Town population = 4,000.
2. The town has a fire station, police station, one hospital, one elementary school, one middle school, one senior high school, three grocery stores, an old downtown area where there are various stores and shops and a movie house.
3. There are two churches, each with a basement. All the schools have basements, but they are damp and musty.
4. The lake area is quite extensive. Numerous cabins are without telephone service. The area is full of tent campers and retired couples. One summer camp houses several hundred children and is located near the lake. The one telephone that services the camp is located at the main desk at the entrance of the camp's 75-acre facility.
5. The town airport has five single-engine planes, each with a four passenger limit, and two helicopters, each with the capacity for six people. The sides have no doors.
6. The closest town is 35 miles away.
7. The town has one radio station but no television station.
8. The local industrial plant is a paper mill. It has few vehicles to help service a mass evacuation. However, since the company president wants to generate good-will with the town, he would be a likely person to help organize his workers in an emergency evacuation plan.
9. There are five stop lights in town.
10. The police station has 10 squad cars with radios.
11. The fire station has two trucks, but they are used for care situations, not evacuation.
12. This is a bedroom community with 60 percent of the population traveling elsewhere for jobs during the week. Emergencies do not always happen on the weekend.
13. A prison is located 10 miles north of town and houses 200 inmates. They are sentenced to life in prison, with records of dangerous crimes to humanity.

Clarksville

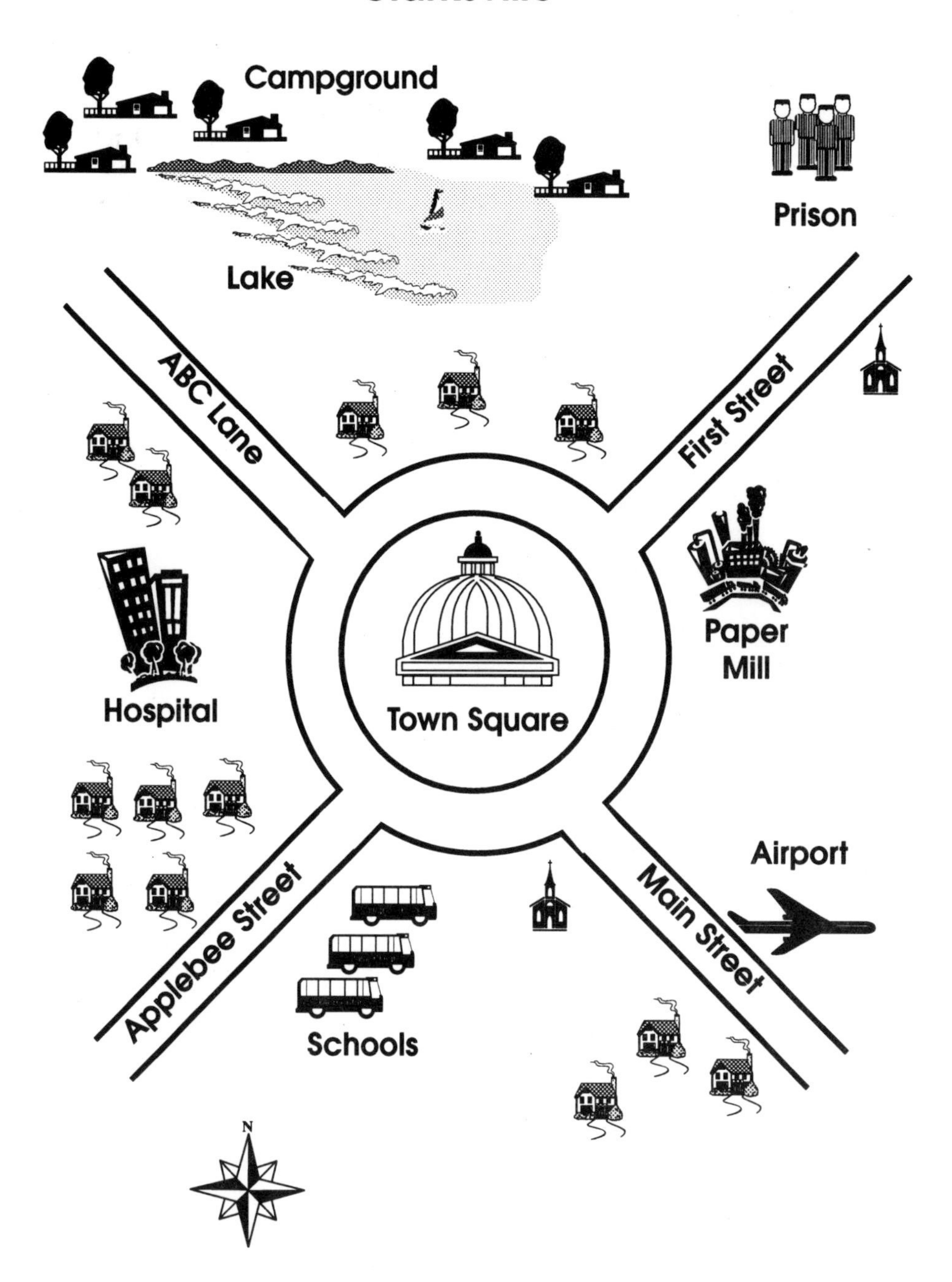

Questions to Answer

Perhaps these questions can be divided up among the council, then when everyone has answered their particular part, the council can review the whole plan. In conclusion, the reporter must have a good copy to hand in for EPs. The mayor of each group will sum up the conclusion for presentation to the whole class.

1. Look at the map of your town and decide on evacuation routes. Which streets would be most appropriate for evacuation and which ones would cause a bottleneck?

2. Where can you store food and water in case of a prolonged stay?

3. What agencies can help in emergencies? What can they do?

4. Who will be responsible for warning the schools, hospitals, residents, and working people? How will this be done with the least amount of panic ensuing?

5. How many planes are available and how many passengers can they take?

6. Is the wind a problem in a toxic waste condition? How do you judge the time you will have before the deadly fumes arrive?

7. Does the weather make any difference in an evacuation plan?

8. Does the time of day make a difference with evacuation routes?

9. How can the industry in the area help?

10. Will you need ambulances? Police? What specifically will they do?

11. What about people like the sick and the elderly who have no transportation?

12. Are there any buildings which might have basements? Can you use them?

14. Who gets out first?

15. What about vacationers in the area who might not be listening to their radios? After all, this is a resort town where visitors come to get away from telephones, television, and radio.

16. Who will you place in charge of the evacuation, the food, the announcements, the school children, the elderly, the vacationers, the sick, and the burn victims?

17. Are there possible refuge spots in town or will you try to get everyone out? Do you compromise, and for what reasons would you consider a compromise?

18. Do you ignore the danger to your family and allow them to be contacted in the manner planned for the rest of the citizens, or do you contact them first and risk a family member panicking and telling the wrong person, who in turn causes a panic?

19. Is there anyone who should be removed from the danger area first?

20. How will you transport maximum security prisoners? Must you plan for a safe haven for as well as from them? Do you divert police and fire fighters from their duty of evacuating citizens to instead monitor prisoners? Many people will be just as concerned about the threat created by mobile prisoners as much as the risk pertaining to a toxic waste spill.

CAUSE AND EFFECT STATEMENTS

CAUSE AND EFFECT #1
Burning of the Rain Forest

A thunderstorm moved across an area of the Amazon Rain Forest. As soon as the rain stopped, clouds of moisture began to rise from the trees to form new rain clouds that moved west, driven by wind, where they provided the water for new rain.

Any interruption of this natural process could have a magnified effect. When large areas of rain forests are burned, the amount of rainfall recycled to adjacent areas is sharply reduced, depriving those areas of rain they need to thrive or even maintain growth.

The effect is a drought cycle—killing more trees thus reducing rainfall and accelerating the death of whole forests. Furthermore, when the canopy of overhead leaves is removed, the sudden warming of the forest floor leads to the production of increased quantities of carbon dioxide. The massive increase in the number of dead tree trunks and branches leads to a population explosion of termites, which also increases methane thus reducing the Ozone layer.

CAUSE AND EFFECT #2
Using Pesticides

Pesticides often leave the most resistant pests behind. Then, when the resistant pests multiply, larger quantities of pesticides are used in an effort to kill them. The process again is repeated. As an effect, soon enormous quantities of pesticides are sprayed on the crops to kill insects—only now the strains of pests are almost impossible to kill and the neighboring communities are complaining of cancer in their children.

CAUSE AND EFFECT #3
Altering Habitats

At least one species of tree-dwelling ape was able to adapt to the disappearing of its forest habitat by learning to forage on the ground and walk on two legs. Thus he used his legs as arms, which had initially evolved to grasp tree limbs, free to carry food and objects.

After reading the Cause and Effect statements, brainstorm and write your own for extra EPs. Think of the problems you have encountered along this simulation to decide upon an issue. Use the space provided to write your statement.

Art Project Suggestions

1. Draw or paint a conceptual composition (a story-telling or action picture that attempts to display a concept). For example, "WHAT IS INDUSTRY DOING TO US?" Draw a composition with giant industrialized machines and equipment, pipes, etc. Run portions off the paper. Suggested media: ink and tempera resist.

2. Draw a series of pictures or a mural to show how the Earth has changed over the years as a result of industrialization.

3. Draw giant insects taking over the Earth as a result of pesticides.

4. Design a billboard making America more aware of the problem of nuclear spills, toxic waste, and dumping grounds. Use bright colors and compose a catchy slogan to go along with your idea.

5. Draw a lush setting as if it were the "Garden of Eden," except that part of it shows the beginning of decay, defoliation, or pesticides.

DILEMMA STATEMENTS

DILEMMA #1
Paper or Plastic?

When you go to the grocery store, you are asked the above question. Think carefully a moment before glibly answering. Which one is really best?

It is true that paper necessitates the continuation of logging our decreasing number of forests, to say nothing about the natural habitat of many animals. Do we need so much paper or wood? Do we have to build our houses out of wood?

On the other hand, plastic is difficult to dispose of, taking too long to disintegrate, and blocking drainage areas. Can you recycle plastic? Is there an answer?

DILEMMA #2
Disposable or Cloth?

When you have a baby, you must think about how you will not become part of the problem, but a part of the solution. Pollution is a serious threat to our environment.

Will you use cloth diapers and wash them in a washing machine, thus using tons of hot water and detergent that seeps into the drainage system and pollutes our rivers?

Or will you think that using plastic, throw away diapers are best? After all, you certainly don't wash them, you merely throw them away. But what happens then?

Again the dilemma of plastic arises. Plastic is not bio-degradable, is it? Perhaps the question is not which is the best, but rather which solution is the least worse? Do you have a solution to these or any other dilemmas that face our planet today?

After reading the dilemma statements, choose one and write and advertising campaign in favor of one side. Try to think of a slogan for your campaign. Use the space provided to brainstorm ideas.

FINAL SAY GROUP PROJECT INSTRUCTIONS

Think about what you have learned during this simulation project. Have you come to some personal opinions of your own?

Instructions: Each participant contributes a minimum one-page report telling what they have learned so far about toxic wastes, DDT, the Ozone layer, specific problems that face us, etc. You may use your notes.

Remember the art projects, the dilemmas and the cause and effect situations. The reporter is responsible for writing a final rendition of the report, including thoughts and opinions from each member of the team. (Make a notation in the margin beside each contribution indicating who was responsible for the work.) Participation must be equal.

Report Mechanics

1. Write an opening statement relating what ecological problems you intend to address, then proceed to write several paragraphs providing facts, feelings, problems and solutions.
2. Conclude in a summation paragraph which provides a projection for the future.
3. Turn in for a grade and EPs.
4. After the teacher returns the work, the recorder tallies up all EPs and the mayors move push pins into position for the last time.
5. Winners are announced.

MATERIALS FOR PHOTOCOPYING
NOT CONTAINED IN STUDENT PACKET

Answers to Ecological Questions
(given to mayors of each group after the ecological test is taken)

1. False. The warmest year on record was 1991. Hypothesis for this warming trend include the fact that huge tracts of forest lands that normally would absorb green-house gases are disappearing. Also, the widespread use of fossil fuels such as the gasoline that we use in our cars, and coal fired utilities and manufacturing plants are causing carbon-based gases to build up in the Earth's atmosphere. This may be trapping heat from the sun's rays and warming the planet.

2. False. CFCs are no longer permitted in most aerosols.

3. C. Fish in New England streams. Acid rain is the name of a growing pollution problem in the United States. It is caused by emission of sulfur dioxide, nitrogen oxide and volatile organic compounds from the burning of fossil fuels. Once airborne, these pollutants are carried eastward by the nation's prevailing winds to the Eastern seaboard, usually in high-altitude areas. Much of these are deposited by rain and snow where they make their way into the lakes and streams of New England.

4. True. Half of earth's species live in rain forests, yet despite their importance, the rain forests are being destroyed at a rate of 41 million acres per year. It is estimated that a quarter of all animal, insect and plant life will be extinct in 50 years, and that all of the Earth's rain forests will be gone in the space of 90 years. The incentive for the destruction of rain forests is primarily financial. Tropical forests, such as in the Amazon area of Brazil, are being cut and burned to harvest drug-related materials and lumber, and to create more grazing land for beef cattle to be sold on the world market. The problem with much of the beef cattle is that it is imported and not indigenous to that area of the world. Therefore, the cattle die out more quickly. But before they do they trample out the crops and pack down the earth, leaving little land to be restored in the coming years.

5. B. Loss of wilderness areas in which to live. Habitat loss is the main problem for ducks and many other animals.

6. C. More than 10 times as much. The annual motor oil dump is 190 million gallons compared with the Exxon Valdez spill of 11 million gallons. Motor oil dumping begins at home where people change their oil then dump it into the sewers where it eventually makes its way into the public water system.

7. False. Drinking water is still at risk. Waterborne diseases are at a higher level now than ever before.

8. B. Metals. Steel is the most widely recycled material.

9. False.